A Search for Human Influences on the Thermal Structure of the Atmosphere

A Search
for Human Influences
on the Thermal Structure
of the Atmosphere

ALLISON LINDQUIST

WINNER OF THE 2021 LIL BOOK AWARD
FROM TELL TELL POETRY

Copyright © 2021 Allison Lindquist

allison@teamlindquist.net
www.allisonlindquist.com

ISBN 978-0-578-30281-2

Edited and designed by Tell Tell Poetry

Printed in the United States of America

First Printing, 2021

Dedicated to the pinecones of one
remarkable sugar pine tree in the
Klamath Mountains of southern Oregon.

Contents

Acknowledgments

To Mr. Harper, Mrs. Letter, Dr. Chase, and Professors Rappleye, Seaton, and Pablo—the mentors who have molded me: I hope you know how utterly grateful I am.

To all the writers I coexisted with during college: Thank you for showing me your words and for helping me to hone my thoughts into art.

To Brodie and Becky, the best friends I will never deserve: I know you don't care about poetry, but I appreciate you trying!

To Justin, who told me I should write a book: Here it is!

Finally, to my new love: You have held such sacred space for my accomplishments and I deeply treasure your quiet excitement.

A Search for Human Influences on the Thermal Structure of the Atmosphere

Brainstem

Surface Tension

I am the drops of winter rain
weighing down the boughs of thirsty trees.

I am circles of light,
a magnifying glass of condensation,
and the changing of seasons.

Can you listen to me?
I am an extension of your beating body.

Gravity works on you still.
I merely let you see the dampened effects.
Your deep sleep upon my arrival is a magnification
of your inner workings.

You will begin to feel the pulse of blood in your ears
and I tell you—it's always been there.

But my quiet music against the cabin wall
reminds you for the first time
of your mother's heartbeat pulsing in the womb.

She calls to you through my refrain
of water and storms

and unadulterated lifeblood,
sighing into gravity's arms
with each heavy drop
against your windowpane—

listen closely.

Tide Pool

brush your tentacles over a scar on his cheek
paint a map of his face in your thoughts
if you murmur he will press his fingers against leaks
all sound is food since it holds our mass taut

paint the map of his face in your thoughts
the flood rains demand all our magnetized ideas
all sound is food since it holds our mass taut
pools of immunity create the panacea

the flood rains demand all our magnetized ideas
clasping the empty shell of a hermit crab world
a pool of immunity has created a new panacea
tide-pool eyes claim your body unfurled

The Effects of Observation on the Behavior of *Corixidae hemiptera*

I am an underwater bug.
Will you name me?

I slice the water, gliding powerfully.
My inch-by-inch movement seems minute to you.

When I break the surface and dive into the silt,
sifting pebbles for food,

you observe me with unabashed interest.
Others catch a glimpse and immediately forget me.

You stay— You watch me glisten in the shadows for hours,
memorizing the shape of my movement,

the dark gold pattern on my back.
If you catch me, will you show me to the world?

I do not see you move, for you do not wish to disturb my work.
Your above-water realm is so foreign to me,

even as I am a part of it.
Watch me closely.

I will show you
it is not a hole dug for the chill.

Your palms slap the water, reverberating with ripples.
I am frozen for a moment, overwhelmed.

I search urgently for signs of winter,
which come without fail.

You stretch to reach me with your fingertips,
but holding me will not keep me safe,

will not help you to memorize me.
Join the rest of the settling pond,

searching for a change of seasons,
searching for my name.

Flesh Traffic

Her scarf is the red glow of *Open* signs
against the wet pavement.
The lust flows so deeply here
it is stained in oily puddles
onto the sodden streets.

All the while, she walks—head down
against the fleshy storm—in the red-light districts.
No one knows where they came from:
the women seeking company.
Just as no one knows why they
are bathed in red light.

Do we blame Rahab
with her scarlet rope advertisement?
Or is it the sailors coming into shore,
leaving their deep-glowing lanterns
on the red doorstep?

They watch her walk
red-eyed into the warm interior
that is humid and heavy with desire,
pulling the scarf closer as she is inspected
with a bloodshot gaze.

> *Anatomical intensity in the wet streets,*
> *psychological intensity in the wet sheets.*

Striving for another
embroidered night of great feats
and conquerors holding the youngest close.
Some things can be left unsaid

until the Revelation song,
singing of the return of a horse:

> *A fiery red one. Its rider was given power to take peace*
> *from the earth and to make people kill each other.*

She is nothing but raw meat
for the hungry appetite of a rust-red sword,
held captive by the thundering hooves
of the return of a Christ.

> For sale with an expiration date
> painted on neon signs: *best if used.*

Somehow, she was like the others;
and they were the color of a newborn,
where shrill red equates
to the beginnings of a new life.

Existence is picking through her body.
Each puncture harbors a scab caked
with the thick red liquid
of broken needles and microscope slides—
the liquid of tarnished magnetism
and humanity.

Concerning the Application of American Patriotism in Daily Life Following the Loss of US Space Exploration Funding

It so happens that when you categorize them all as a conspiracy,
you've got to cram them deeply into a particular variation
of the timeline of the moon landing.

Just stick a flag in it—a cacophony of crying events and silent stares
and their orchestral call for America the Great and Beautiful

and a chessboard of veins and crevices tied up
so that more wrinkles are more ropes,
each thread in turn buffeted by exhaust pipes and politics.

What is it that you were outraged by?
The typewriter of anatomy and I forget what else.

What did people buy and sell before oxygen?
Perhaps men—the first to rendezvous in time. They named it
the Age of Exploration in which nature

told you its secrets would be revealed
only when you explore typography on your shower door,

Maybe equations for the capsulized ecosystem (or was it
an explosion?). Confetti inside a mine shaft won't do any good,
but wear that party hat to the press conference!

The age of cosmic discovery had to be dragged out back,
beaten by schoolyard thugs pledging allegiance to an imaginary
    contrail.

Panic Hole

You can see anxiety
on my skin: layers
bit off my lips and chunks
from inside my cheek.

Is that cannibalism?
If I pick enough skin
off my face and chew
the inside, will I make a tunnel

all the way through?
A new face-hole, something
people will examine
alongside the nose holes,

the eye holes, the mouth
hole. The panic hole:
a symptom of self-digestion.
My anxiety is like estimating

the amount of pasta wrong:
too much starch and not enough
sauce and if I had a hole in my cheek,
would a noodle slide back out?

PhD

A *terminal degree* sounds like an illness—
the cancer of academia,

institutional mutations
fed by diatomaceous earth,

shredding houseflies
with perfectly crafted

glass-shard dirt.

Death by Two Hundred Apple Seeds

after Alfred Sisley's *Nature Morte aux Pommes*

One apple oxidizing beyond recognition, cut
with Sisley's autonomy, steel knife laid inert
beside three brothers: all containing amygdalin
in their brown velvet seeds & Mother said to never
eat them, lest a tree grow out of one's stomach
or cyanide release into human digestive enzymes,
which is a natural death, they say, of acute toxicity,

principally deemed as more elegant & wholesome
than expiry painted out of linseed oil & some Indian
mineral pigment. These four were the only bushel
with leaves still pressed against their taut bodies
& they caught my eye as enzymes catch
free cyanide molecules & render them harmless
once combined with sulfur—passed safely through urine.

Providence Sent Me a Fading Damselfly

I am not departed
but merely in the process of dying.
Only my wings are worth saving.
The vibrating of my capillaries
calls the wind into submission.

Why won't you look me in the eye?
Would you see the reflection of the forest
multiplied ten-thousandfold
within the geometry of my vision?
See the ecosystem erupt prismatic?

I am more a part of you than you realize.
I come from the deep red pine boughs
and the silent river stands,
where the reflection of tree is more real
than trees themselves.

It was in water I was born—
nymph to sighing nymph—
and to water I will return.
For now, I grasp at your shoulder,
until the rapids' echo calls me home.

Sonnet for Sam

Quiet soul, shall I pray thee sleep in peace
or should a dark and dang'rous mind take flight

to land among discarded sentences
where thou might learn to walk on the moon?

Slowly pace thee around cast-off veins,
empty capillaries of the gutter.

Thine eyes crease with the light of salt-rimmed skies.
Carry the bright weight of thy belongings

next to manila canvas and a poem
that talks of the ashes stuck in thy teeth,

a grimy past of overpainted bricks,
and the deep grudge of hypoxic jawbones.

Replay thee the green glow of lightning bugs?
To highlight thy red scabs on bruised knuckles.

Cerebellum

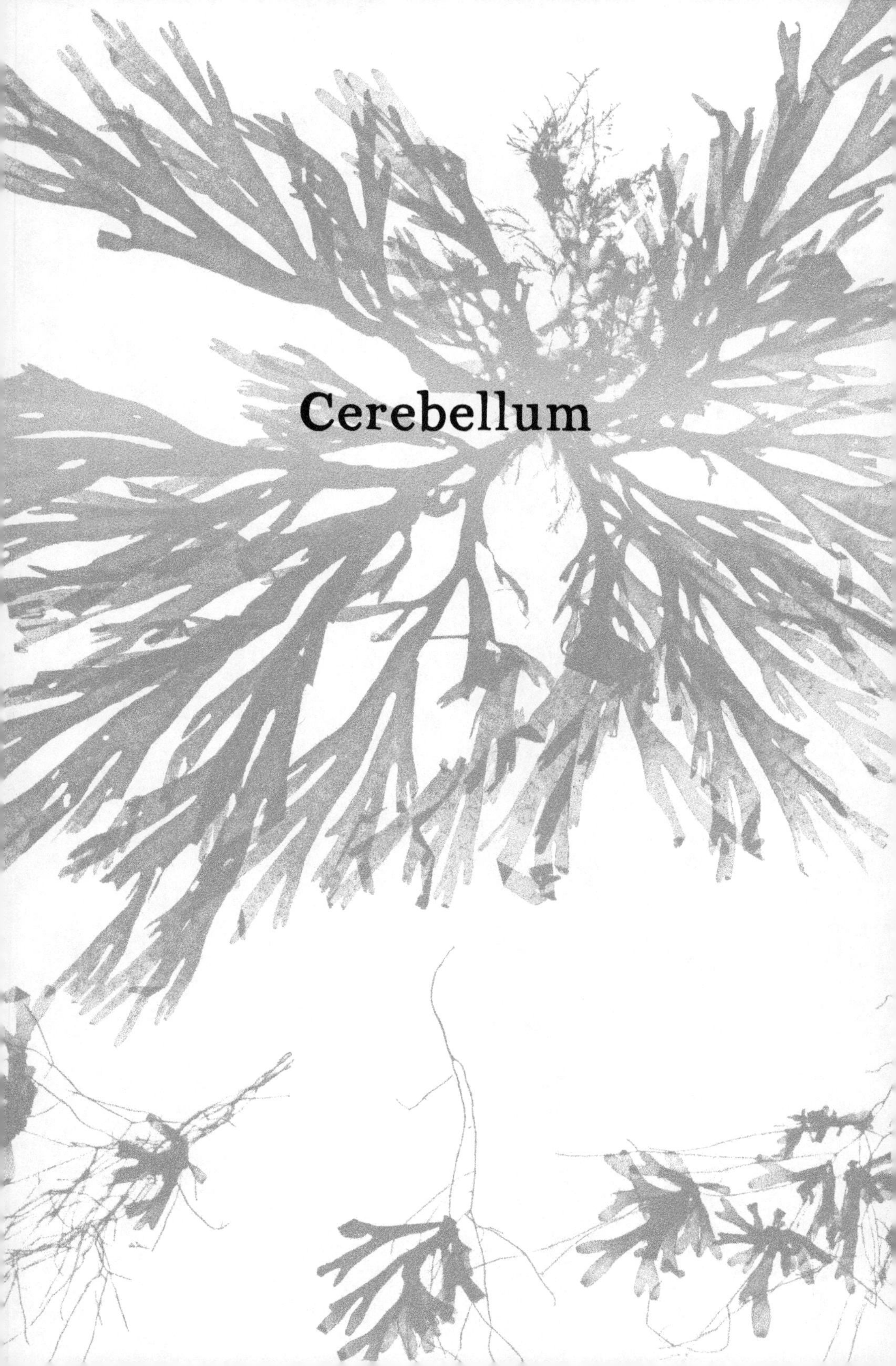

Self-Portrait as Sporophyte

I am the fern,
growing where the forest floor
is no longer ground

but a moss-covered rock
three feet up and there is no difference
as I unfurl, resolute and spreading.

I am the victory cry of a northwest rainforest,
displaying my fronds in humid, dappled sunlight,
where moisture is more common than land.

I, too, will become my brother.
A beautiful death turns my spike leaves
brown and golden

through the bursting of my vine-toothed veins.
I become a mottled curl of memory,
smelling so strongly of home

you can't help but wonder
if you will look this earthen, too,
when you begin to rot.

But I had the audacity to find my footing
far above the leaf layers of the ground.
If you had settled your cheek against my fronds,

into my arched arms,
then beautiful deaths would have been reserved
for those who begin their lives in fiddlehead spiral.

The imperfections make the shadows
beautiful and my death worth framing.
I was waiting for your embrace.

Spatial Awareness

> *When you look up in some forests, you'll see*
> *"a network of cracks formed by gaps between*
> *the outermost edges of tree branches."*
> —James McDonald

The phenomenon of crown
shyness reminds me of living with another,
forcing my neighbor into patterns

that maximize resource collection
and minimize conflict.
If a tree-fight lasts a thousand years, in just
a few tree rings, the wood is split by the sap
of nations collecting growth to scorch

their talk of war and diplomatic meetings,
of bud versus bud and the friction of twigs
jostling to become branches and so on and so forth,
sapping energy.

Like when winter is in the tree canopy,
intimately growing in a pea-soup fog
that floods gaps between trees—
a whole and impermeable forest.

If we encircled the globe with a wall,
would everything be indoors?

Conceptual Red

It's no wonder that we
claim people without sight
are disabled when we say,
I see, to understand not only
vision but concept and emotion.

But what the hell does it matter
if beard stubble and a tulip grasped
in air and in thalamus are coded
differently into existence?
Does it matter whether your red is my red,
if they are both the concepts of red

such that we can understand we are relating
to the other in embarrassment
and warmth? It's like pigeon-
versus people-focused architecture—
pigeon spikes reduce bird shit

apparently, but they also reduce
aesthetic pleasure derived from
looking up at Gothic cathedrals.
The sensation of protection
removes the sensation
of perfection.

We're in the Season of Justice and All That

I'm uninspired and afraid for the bleached corals
and the cancer cure that is in an extinct
species of snake on an island off Brazil
that we won't ever discover because we're too damn
caught up in reading the President's tweets
and screwing the next Tinder date
in the same park we played in as kids,
back when there were three hundred more species
of ash tree than now because the planet
is almost as depressed as the elderly gentleman
showing me pinot noir in Meijer.

Ceiling Invocation

What is my supplication?
Of asbestos, perhaps,
or speckled plaster.

I lay immobile with a prayer
to win my fight for importance
in the tedious jigsaw puzzle
made of mineral fiber ceiling tiles.

My mouth is opened wide
as if to catch raindrops
inside the office building
coated with noise-reducing
industrial contracts.

Lips stretched tight,
the dentist scratches my teeth.
Without words,
I am vulnerable in the chair,
so I count the tiles above.

I hide in my mechanical womb,
cradled by leather warmed by my skin,
as I am laid static against the power
of three hundred and seven
Radar 2 ft. x 2 ft. lay-in tiles.

Theory of Chair

Four points to ground
and flat planes plastered onto
ass cheeks of the masses.
But every chair, according

to Plato, is an imitation of the real—
a concept, formed by experience,
stacked on top of experience
in the essence of things.

I wonder what Plato
would say about plagiarism
if everything is an imitation
and nothing is the original

or even real, for that matter.
It's probably negligible if you
sit on an idea, thighs stuck
to fabric of your own fabrication.

Footnote from the Leaf Found in My Nature Writing Journal after Leaving Oregon

I was the skeleton of an alder
pushed against the mudstream rock;
you plucked me from the strong grip of the water.
A sifting of moss and debris
to expose my startling anatomy—

a cadaver in your hands.
An autopsy display of spread veins,
you were astonished by my intricacy.
The webbing somehow remained,
despite the river's constant wash and press

squeezing the color and matter from my bones.
Drying against your palm,
you held me up to your eyes
like a smokescreen for your vision—
an embrace of visual geometry

that turned to a blur with my spine as a veil.
The product of sun and loam,
the product of a snowmelt current,
the structure of individual cells was left,
bared for winter.

You hold an abstraction of the natural world.
Fold me carefully back into your book
and, in complete stillness,
stomata sink into the thicket of letters.

Remember me as I am, not merely as I was.

A Search for Human Influences on the Thermal Structure of the Atmosphere   25

I Attend a Seminar on Karma

to give myself a moral test
and settle into a cross-kneed heap
on the wood floor, saturated

with hydrogen and oxygen
and carbon-based cries of dust.
My ego is the King of Pop

and requires the psychotherapy written
by an ancient Greek slave's melody
of call and response.

Sound is water vapor,
not in matter but experience.
I, too, am drawn, lightning-bound—

an idea of circular reincarnation
closer to Communion wine
than the *chimarrão* I am sipping.

With steam along liquid,
I wonder if Nirvana fans believe
I'll come back as a carnivorous beetle.

Partner Eleven

I steal the clouds, the moon, and we sway—
you with me—an ally to the day of my birth.
We harmonize the flow of our celestial spaces.
Only when you explore layers of sediment—
a gravitational timepiece settling bones into fuel—
can we keep a careful eye on the pickerel frogs
looking through a lock that requires three people
and a pair of chopsticks—all shriveled under the sun
like the before-product of Sunday Communion,
while women shout expletives without batting
an eyelash and we brush stucco with belly and cement
with ribs. Alone, it is only a noise in the multitude
and is not to be forced. Don't you remember?
The bird lives in your room.

Cerebrum

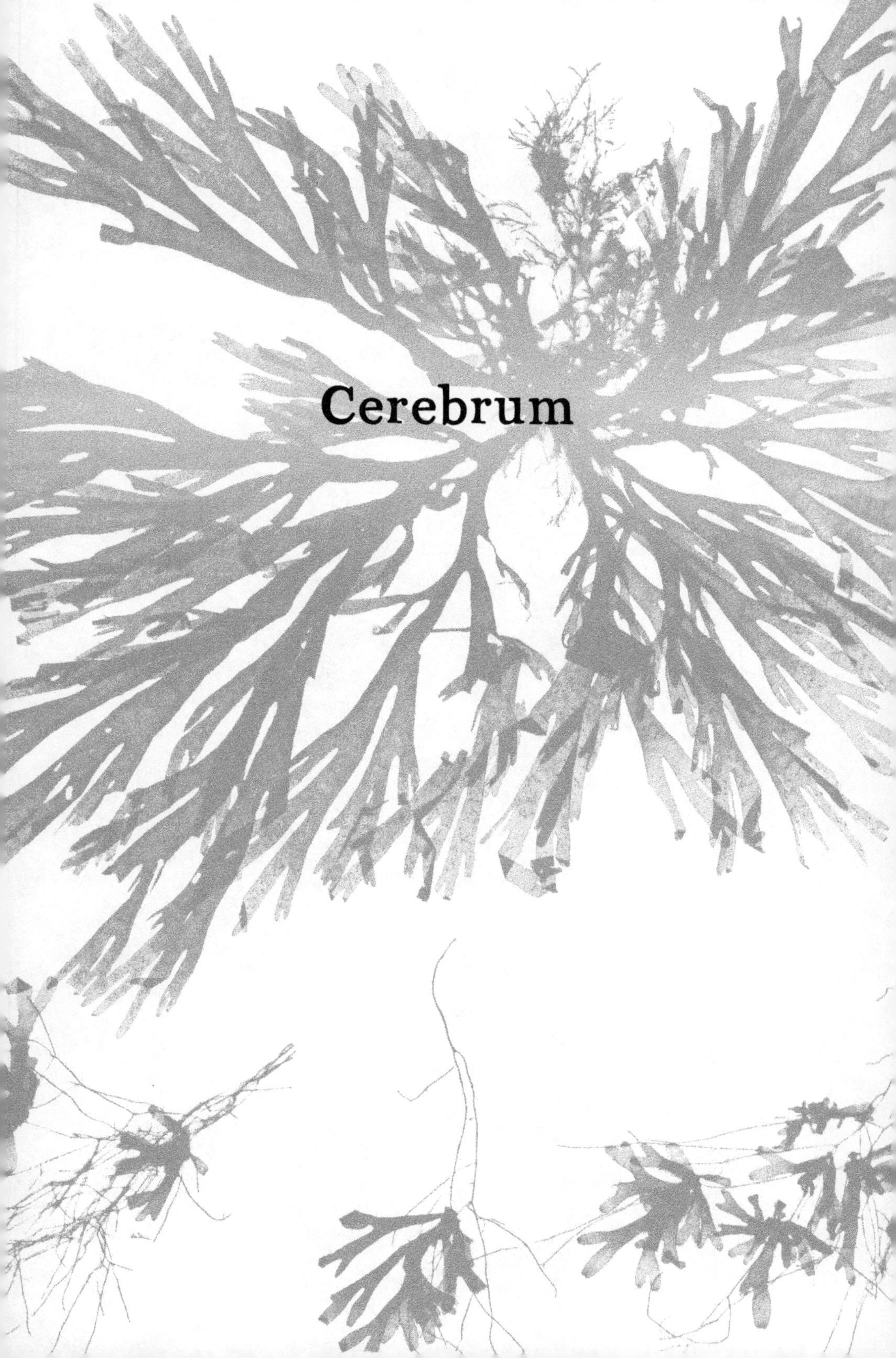

Enthalpy of Vaporization

I am the steam wafting,
ephemeral, into the night air
as you cry out
to the rainbow-shadows of the moon.

I am the haze you wipe from your lips
in a harsh word punctuated by a kiss,
a lung-warmed breath that quickly
fades to nothing.

Even in the opposite of existence,
is my memory enough
to remind you of the dust of your being?
Ash to fallen ash

as vapor covers your face in a momentary white.
I am the prickle in the depth of your diaphragm;
but your body is warm
and an exhale means the beginning of gas exchange.

Life in nodes of carbon and oxygen,
turning into your breath and life and death between.
I am here, but only long enough for you to remember
the pain when I am gone

but not forgotten. No, I am remembered
in pharaoh's tombs and
coronations and the first cry
of a child, all borne

in a sun-held curl of breath.

Grandmother

all our collected seashells are drying
salt-skin on paper towels and we mix
colors like turmeric
and cranberry juice singing

I love you a bushel and a peck
and we drip watercolors
slowly and carefully
following the seaworn ridges

she guides my untrained hands
and a barrel and a heap
and *I'm singing in my sleep*
with fading memories of you

A Search for Human Influences on the Thermal Structure of the Atmosphere

—B.D. Santer et al., *Nature*, vol. 382, July 4, 1996.

rest your eyelids
in the shallow of her back
hold in your thoughts
       until they flow

out your ears
the portal to the soul
is just a sound
       more than mass

cringe against the rising tide
lust mixed with the weight
of the world where peace
       is shards of a clay pot

discovered in an 1887
Honduran archeological dig
and she now dozes
       under a museum

velvet exhibitions
of dust and stored maps
of a free radical that no longer
       hugs her curves

Richter Scale

She was a glass wrecking ball,
the archetype of a scientist reflected in an Erlenmeyer flask,
a being formed carelessly like the cracked
scraps of clay on the edge of the potter's wheel.

She speaks to me in undulations—formless master,
where nourishment has no price but existence.
Doomed for empty destruction, crying out to hardened clouds,
You have startled my drowsy heart.

Made for so much more than a pair of languid lips in the void of space.
Stitched with ecstasy—joyous insanity.
Insatiable curiosity born out of the destruction
of an intimacy no louder than a 2 a.m. mist.

Ethos taught her to tame me—holding centuries
before her to mold me, whip me, thrash me
into the submission of cups and concrete pools and metal sidewalks.
But I hush her and sweep my edges to the spectacle of her silence.

With storms and potholes—nuisances of survival—
this heart-sting is why she writes the present. But compassion
for the future is why she stumbles inside yawning lava tubes.
Show me a pattern deep in the earth.

Haiku from Medial Brain

a synaptic cleft
mediates the inner world
in background music

the hippocampus
is a seahorse in silence
hugging ventricles

sex and aggression
ten percent perspiration
to fight the urges

amid the left brain
choose another damn circuit
this one is in use

neurite disruption
the infection is spreading
momma I'm tired

increased viral load
he was found dead on the bridge
nature or nurture

clouds must wax and wane
to the shadow of the moon's
metamorphosis

Chronicle of Poor Richard

You may delay, but time will not.
—Benjamin Franklin

I remember the colonies as a child—
digging in their dirt, searching for ideas to print.
Each letter was huddled under the roots of a winter tree,
waiting for me to believe in their weight:
fifteen and a half grams of pure possibility.

Out of one hand come the visions of many—
a declaration of freedom, a forecast of crops,
planted in the New World.

I weighed the kite down with the keys of knowledge,
but it sailed into the clouded atmosphere of democracy,
straining past the British lantern's silent warning.
A thunderstorm prepared with electric fire
sealed in a glass canning jar till summer.

What use are blueprints for an impending future,
if the only way to view them is through soiled spectacles?

My bedtime tales are filled with British buildings owned by oculists.
They rate my sight—tiny numbers printed neatly into the almanac.
An excuse to bring glasses *to mend you*, they said.
Wire-rimmed frames hanging like a frozen handkerchief, from
    one grimy hand.
I am etched into commercial plaster: *the benefits of a life full of clarity.*

Cartography is irrelevant now that West has become East.
Don't you know? The world is a circle—
Don't waste your square paper.

Someday another child will see the world as I did,
shaping the clouds with the bubbling innocence of their mind.
He sees a rabbit where I once saw the Spirit of America,
breathing out its form—a serpent singing *unite or die*
until the westerly wind hurries us along
the cobblestone waltz to the past.

Is the ink of the pen still mightier than the residue of gunpowder
while fighting a revolution in the crumpled bedsheets of time?

Study on the Growth Rate
of *Letharia vulpina*

Lichen clumped on tree limb
yesterday in the golden wheatgrass,
clustered together for warmth.

Feeding and fed
in miniature, a forest on a branch.

Darker strands are garden leaves,
the lighter like desert coral—
all a shade of barely-green,

as though a painter forgot
whether he was creating ground or sky.

Even the black-hair moss
hints epigraphs of green
to tree-home origins.

Rise to the sun in curves
we do not think to notice—

dry and strong growth
that is the hue of hope,
like falling into a lover's

sea-storm eyes.
This incubation is a song

of wind and sunshine,
all wrapped up in one arched spine.
And so, sing to us

and we sing back
in harmony with kin.

A love song whispers history by name,
each verse an inch marking
wet fallen pine.

Abecedarian for Jesus

A poet once told me something he discovered about Mary's
Bastard son, Jesus, in the ancient Hebrew times. The poet was
Considering the state of the union and of injustice, and holding
Disgust at arm's length through an iPhone screen. An
Earnest student asked him: *How do you*
Feel about using your privilege to address inequality? And our
Good poet paused for a moment and spoke quietly:
Have you ever considered the story of Jesus and the adulteress?
In fact, Jesus uses his privilege to dissipate the mob
Just by crouching down and drawing in the dirt.
Kingdom come in opportunity, I suppose, because only
Lucky men could read in those days. Jesus stripped the
Mob of its power when they began to ask each other which
Names might be taking shape under his fingers.
Or perhaps Jesus could just see that the crowd needed something to
Pull their eyes from the pain of another and look back at their own
Quietly grotesque human nature.
Really that might have been what the gospel is all about. Maybe we
Shouldn't be afraid of the talents that we have been given, even if
They are just writings in the dust. Privilege
Under God is remembering to convict one fewer scapegoat for
Violating the rights of those who are truly
Worthy. Unless someone believes that all individuals are
Xenoliths—where every man is an island—and can be found
Yelling that Cain killing Abel was nothing less than orthodox
Zeal in the absence of social accountability.

Pocket Lint Number Seven

Do you remember catching crawdads with hot dogs?
Back when the buzz of power lines
was almost indistinguishable
from the grasshoppers

who painted my hands with their blood.
Those copper droplets, now my favorite color.
And me? I live where the power lines end.
Where my joints tingle from clutching
bike handlebars too tightly.

I ask you to draw me—like one of your French
fries, forgotten beneath the seat of your car,
collecting pocket lint and memories.
Caught between the plastic plains
and the evergreen wall of the Rockies,

where the cattails were taller
than my tangled hair and the wild
takes over again. Only giving space
to the windmills and the occasional coyote,
fighting fire ants and yucca's spiked growth.

I ask you to draw me—brushing helicopter leaves
down from the thick branches,
simply to make you smile until our next water-fight,
when life smells like chlorine puddles
and sun-bleached plastic.

Those were the sweet times
when moths were friends,
but only until they hatched in our care.
The ladybugs raced the raindrops over the fence.
I'd pretend they'd win for you,

hiding their sopping wings with my superiority.
Lording over our youth,
it was just us two, humming in anticipation
of "not another" shadow puppet show,
somewhere underneath a twig soup fort.

That Night You Got a Haircut

You and I, we share a cigar,
celebrating the arrival
of another night perhaps.

Two friends, you and he, crush my legs—
one on either side—
into the granite of a faded memorial sign,

pressing into my skin the name of someone
who treasured the duck pond and the black oak,
both too quickly overlooked in the green-grey folds of the park.

We sigh tobacco smoke into the earth-laden air,
adding a white film to the water of the pond in Lithia Park,
tinting the dusky night violet.

Drug-store bourbon passed between us three like a talking stick,
each one in turn spilling scraps of mythological origin stories—
the cliché of capitalism and a damaged love life.

We sit discussing structure and antistructure
and our own self-righteous indignation
at being unable to trace the contours of a heart in motion.

We're alone in the park,
hidden beneath the wire-wrapped fields of Kant
and a powder-scarred lyre and a misplaced need to protect.

Does howling the drunken truth
at a park bench mean the words ring less real?
We look down the empty barrel of the bourbon and snuff out the cigar.

Is this the same as an intimate smile with a stranger?
Far too potent to be recorded beyond
the sallow carcass of this night.

Where I Met Joy

I met joy in a wet, greasy alleyway
between crumpled candy wrappers
and soggy paper towels.
She held out her arm like a peddler
asking for spare change or a bus ticket home.

I met joy curled at my feet like a wet dog—
submissive and warm and breathy—
dripping between my toes
and leaning into my bare ankles.
She gazes glossy-eyed
and bright into my shrouded thoughts.

I met joy on a blanket-lined couch
as I held her jaw and rubbed her heart
away from pain. Telling myself it would
all be okay, as she knelt exhausted
over the plastic wastebasket and heaved.

I met joy at the same moment I put a name
to disgust at the litter strewn across streets,
tempting their gold-shine
with window-shopping reflections.

I met her there, behind me
in mottled view of a display case,
in *déjà vu* and conditional survival.
(She has carried a pocketknife for so long,
it's made a chalk ring in her jeans pocket.)

I met joy—but I can't keep her
away from the darkest drawers of my heart.
She rips through the cobwebbed files,
shredding memories to dust.

She tells me:
                    You are not entitled to your pain.
Handing me my crushed thoughts,
sifting through her clenched
fingers into my grimy palm.

Now I know joy—but I can't keep her
out of my grocery lists, stained
with wood grain and hurried pen marks.
The urgency of granola bars
and mouthwash seems derisive
to the intelligence of my youth.

I desire joy the same way I desired him—
servant-hearted in the library of his muscles.
Anatomical pins stuck
in the *vena cava* next to *carotid*
next to the *sternum*,

superior to the missing rib
that all men display proudly
on mantles next to the heads of doe kills.
Desire met joy in conquest,
invading her intimately as we lie
synchronizing our breathing

in opportunity that smells of coffee
grounds and flea-market maps,
yellowing her with knowledge of lust
and the invitation of destinations to come.

I sometimes want to forget about joy,
but I know the truth—
those memories are not sticky notes
I can fold into my pants pocket
to forget about until laundry day.

Still, I cling to joy—sweating in my pocket
next to a graphite love poem
on a Trident gum wrapper—
until she slips out
through the unpatched hole,
taking with her the smudged words.

A wrapper to give away
when she meets someone new.

About the Author

Allison Lindquist grew up outside of Boulder, Colorado. A neuroscientist by trade, she holds a research job in West Michigan. With a background in painting, poetry became her brush of choice when met with the inspiration of a teacher who believed in the power of words. In her first collection of poems, Allison's work explores the complementary nature of science and art.

www.ingramcontent.com/pod-product-compliance
Lightning Source LLC
LaVergne TN
LVHW041237080426
835508LV00011B/1254